THE OPEN MEDIA PAMPHLET SERIES

3

THE OPEN MEDIA PAMPHLET SERIES

Gene Wars

The Politics of Biotechnology

KRISTIN DAWKINS

SEVEN STORIES PRESS / New York

A Seven Stories Press First Edition published in association with Open Media. Open Media Pamphlet Series editors, Greg Ruggiero and Stuart Sahulka.

In the U.K.: Turnaround Publisher Services Ltd., Unit 3, Olympia Trading Estate, Coburg Road, Wood Green, London N22 6TZ U.K.

In Canada: Hushion House, 36 Northline Road, Toronto, Ontario M4B 3E2, Canada

Library of Congress Cataloging-in-Publication Data

Dawkins, Kristin.
Gene wars: the politics of biotechnology / by Kristin Dawkins.
p. cm. — (The Open Media Pamphlet Series)
ISBN 1-888363-48-7
1. Agricultural biotechnology. I. Title. II. Series.
S494.5.B563D39 1997
338.1'62—dc21 96-39684
CIP

Book design by Cindy LaBreacht

Seven Stories Press
632 Broadway, 7th Floor
New York, NY 10012

Printed in the U.S.A.

9 8 7 6 5 4 3 2 1

DEDICATED TO JOHN FLETCHER

of the Federation of Southern Cooperatives / Land Assistance Fund of Albany, Georgia.

I myself have plenty to eat. In fact, I often eat too much and need to watch my diet. But there are millions of people in this country who are malnourished and some are even starving. Anyone who lives in a big city sees it regularly; in rural areas, it may not be as noticeable but it's just as real. According to official U.S. statistics, one out of seven Americans lives in poverty. One out of six—45 million people—use food stamps to keep hunger at bay, while 25 million go to food banks for additional help. As many as a third of these people are children.

With the Welfare Reform Act, our Congress cut $57 billion in help to the poor out of the federal budget. Food stamp benefits are being cut by 20%—down from about 80 cents per meal to just 66 cents. Unemployed adults with no minor children are eligible for food stamps only 3 months out of every three years. Federal aid to families with dependent children—the AFDC program—is eliminated, cutting off a major source of financial support to 13 million people, of which more than 8 million are children. With the cuts in just these two programs—food stamps and AFDC—

poor families will lose about $660 per month. For the average family of three defined as "poor" in the United States, this loss amounts to two thirds of their former monthly income. How will they live on the other $330? Can you imagine living for one month on $330? What if you had two little kids?

Bear in mind that the states, which are supposed to take over the responsibility for providing welfare, are already having their own budgetary problems—and they are under no federal obligation to provide services, even if they do accept federal funds to do the job. On the contrary, the states are limited by the Welfare Reform Act in what they can offer the poor: even the maximum benefit level will be significantly lower than in the past. No one, for example, will be eligible for state welfare for more than 5 years over his or her entire lifetime. A parent whose children are more than one year old must do paid or unpaid work after just 2 years of welfare, and the states can opt to shorten this limit. Furthermore, the states do not have to provide child care or transportation, as was the case under the former federal work requirement.

How can the government of the richest country in the world get away with this monstrous welfare reform?

According to our government's own lawyers, this Welfare Reform Act could have been deemed a violation of human rights if last year's World Food Summit, a United Nations-sponsored meeting of top officials from more than 150 governments around the world, had approved widely-supported language establishing the "Right to Food." The U.S. government was virtually alone in objecting to this concept and succeeded in blocking the rest of the world in recognizing such a right, even on a non-binding basis. The final document simply commits governments to "clarify the content of the right to adequate food" and to "[m]ake every effort" to implement it.

Actually, ever since 1948, the U.S. has been legally and morally obliged to the Universal Declaration of Human Rights, which in Article 25 says (sorry about the gender references; remember, this was written in 1948): "Everyone has the right to a standard of living adequate for the health and well-being of himself and of his family, including food, clothing, housing and medical care and necessary social services..." Why has this country failed to comply with this fundamentally decent mandate? I guess in 1996, given our peculiarly powerful position within the world community, our government for some reason believes it is immune from international commitments—except, that is, trade policy, which just happens to enforce the rights of corporations instead of the rights of human beings.

In the following pages, I will discuss a series of international policies which are shaping the future that our children will inherit. Perhaps most alarming about this future is the likelihood that despite new technologies, the number of people in the world going hungry will actually increase. The genetic engineering of seeds and plants is a major factor, and we will see how a diversified gene pool is crucial to food production—and how corporate control of the gene pool threatens our collective security.

Behind these issues lies the specter of globalization—transnational corporations freely exploiting the resources and consumers of the world while political power shifts to remote international institutions strictly dedicated to commerce. Our challenge is to understand all this, and to develop global systems of political discourse in the public interest. It is my hope that this book will contribute to your understanding of these problems and show a few of the ways in which we citizens can make a difference.

FOOD INSECURITY AFTER NAFTA

Anyone reading the newspaper knows that President Clinton and Presidents Bush and Reagan before him have been devoted to free trade and used every political tool at their disposal to accelerate the negotiation of free trade agreements with reluctant trading partners. Citing text book theories, they pretend that free trade is the only engine for economic growth and that, gee gosh, if our big corporations seek business opportunities everywhere else in the world, somehow it will all trickle down for the benefit of each of us. How this can be true is beyond me, with transnational corporations paying little or no taxes, here or anywhere, and robotization generating gigantic increases in productivity per unit of labor, while poor folk don't get the kind of education they need to compete for the high-tech jobs. But the theory is, what's good for corporations is good for America.

During the fight over the North American Free Trade Agreement (NAFTA) in the U.S. Congress, critics pointed out that the only winners would be those corporations which could afford to move their operations to Mexico, thus improving their bottom line by avoiding labor unions and the federal minimum wage, as well as the costs of complying with U.S. environmental, health, and safety regulations. The result would be job losses here in the U.S. and dislocation in the Mexican economy and culture. Sure enough, by 1996, economists counted at least half a million U.S. jobs lost due to NAFTA; in one re-training program alone, the U.S. Labor Department itself certified 98,645 workers as having been laid off due to NAFTA.

The benefits predicted for Mexico—primarily, new investment—disappeared in December 1995 when the peso, which had been kept artificially high throughout the negotiations and subsequent Mexican election, was sharply devalued—from 3.1 to 7.88 pesos to the dollar. The devaluation meant that costs more than doubled for anyone spending money in the Mexican economy. Investors, therefore, snatched their money and ran to safer currencies like the yen. Meanwhile, anyone trying to save money in Mexico found their assets were suddenly worth less than half what they previously were.

Economic activity came to an abrupt halt. Interest rates soared to 120%, but wages crashed. The Mexican government estimates that 2.8 million Mexicans lost their jobs during the first 2 years after NAFTA, despite the creation of some new jobs where U.S. companies relocated along the border and in free trade zones.

Mexico, an agricultural economy, has 40% of its arable land planted in corn, which is the principal food in the Mexican diet. The Mexicans were told that NAFTA, according to free trade theory, would enable them to buy corn from the U.S. more cheaply than growing it themselves. So the government dropped its subsidies both for corn farmers and for tortilla consumers, which together had ensured adequate levels of domestic production to feed everyone with all the tortillas they could eat. Sure enough, without government support, Mexican production of corn and other basic grains fell by nearly half; 25 million acres went unplanted and by 1995 some 2 million peasant farmers headed towards the already overcrowded cities.

And guess what: in 1996, the U.S. nearly ran out of corn. A late spring and a wet summer threatened a poor harvest and since, in the spirit of the free market, our government had sold off most of our publicly-held reserve stocks, the scarcity caused prices in the futures markets to double and triple. Mexico was out of luck.

Conditions in the dry Northeastern part of the country got so bad that women and children actually hijacked trains carrying U.S. corn to Mexico City and carried it off by the bucket-load and plastic bag-full to feed their communities. Nationally, per capita consumption of corn and other basic foods dropped by about a third. According to Mexico's National Nutrition Institute, nearly 1 out of 5 Mexican children is malnourished. And in Chiapas, where the Zapatistas declared their moral war against political corruption on January 1, 1994, the first day of NAFTA, four out of five people—both children and adults—are malnourished.

Surely NAFTA violates the spirit of Article 25 of the Universal Declaration of Human Rights, and should be evaluated as governments "clarify the content of the right to adequate food" and "[m]ake every effort" to implement it.

THE GATT AGRICULTURE AGREEMENT

The global free trade agreement, called the General Agreement on Trade and Tariffs (GATT) is, like NAFTA, designed to further expand markets for big grain traders by driving out national and local production which can't compete with subsidized exports from the U.S. and Europe. Established in 1948 as part of the process for rebuilding the world economy after World War II, GATT was originally set up to negotiate lower tariffs (which are, basically, taxes on imports) in order to stimulate commerce. Since then, there have been seven "rounds" of negotiations, each of which takes several years to complete. The most recent round, known as the "Uruguay Round," placed new issues on the negotiating table, such as "non-tariff" barriers to trade, which includes environmental policies and economic development strategies. Ultimately, what came out of the Uruguay Round was a plan for the wealthiest countries in the world to use "free trade" ideology as a means of dismantling virtually all regulation of national economies.

And the effect of this raid on foreign agricultural markets was deliberate. U.S. Senator Rudy Boschwitz of Minnesota—a big farming state where I happen to live—wrote in a letter in 1985 to Time magazine: "If we do not lower our farm prices to discourage these developing countries from aiming at self-reliance now, our world-wide competitive position will continue to slide." And in 1986, President Ronald Reagan appointed Daniel Amstutz, a lifetime executive of the

Cargill Company, the biggest grain trader on Earth, to be the United States' chief negotiator of the latest GATT Uruguay Round Agreement on Agriculture. Sure enough, between our Congress and the Uruguay Round GATT negotiations, the U.S. got the world price of grains down so low, our exports could out-compete producers almost everywhere.

The United States presently supplies more than two-thirds of the world's grain. Therefore, when prices change in the United States as a result of our national farm policy, the world price changes, too. For Cargill and other big grain exporters, lower prices mean they can buy cheap and sell dear (which is the goal of any trader), while gaining access to new markets in parts of the world where local prices are higher. For farmers, however, whether in the United States or in the Third World, this trend towards a low price is devastating. In this country, some 5 million families lost their farms in recent decades as Congress responded to the big grain compamies' lobbyists and artificially-lowered prices through a complex system of subsidies and increasingly less secure federal loans. Overseas, hundreds of millions of farmers lost their livelihoods, too, as Cargill and other traders dumped cheapened grain in foreign markets. This trend has been exacerbated by the most recent Uruguay Round of GATT negotiations, precisely as Senator Boschwitz and Daniel Amstutz intended.

The Uruguay Round Agreement, for example, requires countries to import a minimum percentage of all basic foods—even milk, readily available from the village cow. All governments are required to cut back on their national supports for farm production, support which was essential to offset competition from heavily subsidized grain and beef imported from the United States and Europe. Indeed, even though all governments are supposed to cut back on national support, the United States and Europe (which similarly exports subsidized agriculture products) cut themselves a special deal, pegging the subsidy cuts to 1986 levels, which just happened to be their biggest year in history—meaning that even if they complied with the stipulated reductions, they could maintain extremely high levels of subsidized exports.

And the Uruguay Round legalized this unfair competition, called "dumping," when the overseas price of a product is lower than what it costs to produce it at home, allowing the U.S. and Europe to continue dumping great volumes of subsidized foods that drive local farmers around the world out of the marketplace.

The beneficiaries of these subsidies are not the farmers, as many people mistakenly believe, but the giant exporting corporations. In the U.S., millions of farmers have gone bankrupt over the past few decades as our Congress, heavily lobbied by agribusiness, has lowered farm prices year after year. North American farmers get only 4 cents out of every consumer dollar spent on food; for the rest, about 20 percent goes to the government, 21 percent to the retail store, 26 percent to the processers and 29 percent to the brokers, traders and shippers. A farmer must produce and sell 104 pounds of corn to buy a 25-ounce package of corn flakes, or 93 pounds of potatoes to order a 3-ounce dish of potato skins with melted cheese.

Nor do consumers benefit. Consumer food prices rose 36 percent between 1981 and 1987, years in which farm prices in the U.S. fell from near three-quarters to just over half of what it costs farmers to produce the stuff. In 1990, packaged cereal prices increased 6 percent while grain prices fell an average of 11 percent. As for milk, the farmers' price fell about 28 percent while the retail price, what you and I pay, dropped just 3 percent from '90 to '92. And these trends are continuing and will continue to continue, given last year's farm legislation.

Indeed, ever since the 1950s, farmers have been going bankrupt in this country due to prices below the cost of production. Back then, the federal government maintained a low-interest loan program designed to let farmers hold onto their crops until the food companies offered a fair price. Then they'd sell, and be able to pay back the loans with interest at no cost to the government. Last year's legislation eliminated just about all government support for basic grains over the next seven years, so chaotic production in the near future is likely to continue causing unpredictable price swings—like last fall's sudden 50-year high which has already fallen back by more than half. This volatility is good only for speculators in the commodity exchange markets, who can make a buck gambling on the change either way; for farmers and consumers and the net food importing countries—like most in Africa—such price swings are a disaster.

Already, just two years into implementation of the Uruguay Round, the United Nations Food and Agriculture Organization estimates that the Uruguay Round will be responsible for at least $1.4 billion out of a total $10 billion increase in the costs of basic food imports for the poorest countries. Their analysis concludes that both weather and the required reductions in government support have caused price increases; that food aid will remain low because the Agreement no longer permits building up government stocks as a supply control measure; and that the reductions in export subsidies required by the Agreement will structurally change production patterns and thus import supply availability.

If this is the case, surely the Uruguay Round Agriculture Agreement should be considered a violation of the Universal Declaration of Human Rights, and governments should reconsider their commitment to free trade in food as—and I again quote the commitments of last year's World Food Summit—they "[m]ake every effort" to implement the "right to adequate food." And surely the "Marrakesh Decision" should be implemented, too. The "Marrakesh Decision on Measures Concerning the Possible Negative Effects of the Reform Programme on Least-Developed and Net Food-Importing Developing Countries" is a little side agreement of the Uruguay Round that gives the poorest countries of the world a bit of financial aid if they can prove that the Agriculture Agreement is really affecting them adversely.

Fortunately, the whole global Agriculture Agreement is about to be renegotiated, since the U.S. and Europe could only agree to six years of peace in their Uruguay Round feud over global markets. And so, in the remaining years of this century, the rest of the world has an opportunity to bargain for fairer terms of trade, planned production, and stabilizing buffer stocks through a linked system of local and regional grain reserves—all in the interests of food security.

This bargaining must occur primarily in the World Trade Organization, that new institution created during the Uruguay Round of trade talks which has executive, legislative and judicial functions, designed to enforce free trade as if it were the only legitimate basis for global governance. But perhaps the United Nations High Commission on Human Rights and the Food and Agriculture Organization, which sponsored last year's World Food Summit where the right to food was undercut by the United States, can also play a role in building a global consensus that food security and human rights are higher priorities than free trade.

CORPORATE RIGHTS VS. PEOPLES' RIGHTS

Actually, corporate rights are firmly embedded in U.S. law: back in 1886, the Supreme Court simply decreed that corporations are persons, giving them the same constitutional rights as people. The right to free speech, in other words, gives corporations the right to blast our communities with liquor and tobacco ads and the right to turn our electoral system into political patronage for corporate donors. Some activists are seeking to reverse this trend, noting that corporations based in the United States operate according to state charters granting them the right to do business with the will of the people of that state. Delaware, for example, the most lenient state in terms of taxes and corporate liability, hosts more company headquarters than any other state in the U.S. What might the people of Delaware be able to do to enforce some degree of corporate accountability? Some activists are proposing the revocation of these state charters by the people of that state, if the companies don't become more responsible.

In testimony before a committee of the U.S. Congress in 1994, consumer advocate Ralph Nader asserted that the rules of the World Trade Organization (WTO) constitute "a Corporate Bill of Rights." The Uruguay Round agreements, he said, "would strengthen and formalize a world economic government dominated by giant corporations, without a correlative democratic rule of law to hold this economic government accountable."

Instead of sovereign contracting parties who choose to participate or not in any of the 180 or more treaties comprising the earlier GATT system, the WTO has "members" which must agree to each of the agreements of the Uruguay Round in order to participate in the world trading club at all. Self-exclusion is almost unthinkable for governments, especially for developing countries bound to structural adjustment policies, net food-importing dependence, and the "open investment" regimes aggressively promoted in today's integrated economy. Indeed, one of the WTO's explicit goals is "achieving greater coherence in global economic policymaking" with the IMF and the World Bank.

The WTO even assumed the power to require each member to "ensure the conformity of its laws, regulations and administrative proceedings with its obligations as provided" in the Uruguay Round agreement. Under these rules, the WTO can oblige members to enforce the revision of certain national, state and local laws—such as regulations for pesticides, procurement of locally grown produce or nutrition labeling—to minimize their "trade-restrictive" effects or to bring them into compliance with often weaker international standards. For corporate lobbyists, the international arena provides them with a nifty way of evading regulation, and for their supporters in the U.S. government, it gives them a way of gaining new posers. For example, according to the Office of the U.S. Trade Representative, even "Indigenous Tribes" recognized through a century of treaties negotiated according to the U.S. Constitution, will be considered "sub-federal jurisdictions" under the WTO.

With the creation of the WTO, the traditions of GATT itself—flexible negotiations among consenting parties through a series of negotiating rounds—were abolished. In their place, the WTO uses voting and binding dispute resolution procedures backed by economic sanctions to enforce its decisions. Chief among the WTO's characteristics is the legislative and judicial power to address areas that had formerly been strictly national in scope by prefixing the adjective "trade-related" to any issue at all. The WTO is a permanent political body and can, with a three-fourths majority, establish new obligations at any time; under the old GATT, the principle of non-discrimination gave every country an implicit veto toward any new obligations that might be proposed.

Disputes resulting over implementation of the Uruguay Round are decided in secret by panels of trade experts appointed by governments; the rules prohibit the members of a dispute panel from releasing documents and from disclosing their opinions to the public. When a dispute panel judges that a country's trading behavior or domestic law does not comply with the rules of the Uruguay Round agreement, the country may bring an appeal before another panel of appointed appellate judges. Whereas formerly a unanimous opinion of all the GATT contracting parties was required before retaliatory trade sanctions could be imposed to enforce a dispute panel's finding (and there were no appeals and no appellate panel—generally, a losing country would negotiate a satisfactory settlement well before retaliation was proposed), the WTO appellate panel's findings are absolutely binding unless all WTO members—including plaintiffs and defendants—unanimously agree to reject its decision.

A losing country must change the offending practices, laws or administrative procedures within a reasonable period of time. If it fails to do so, the winning country can retaliate by asking the WTO for permission to suspend a certain amount of its trade with the loser. For example, if a country refuses to change a food-related law judged to be unnecessarily stringent, it could lose opportunities to trade in agricultural products with the other country—or pay an equivalent monetary compensation. If this penalty is ineffective, then the winning country can "cross-retaliate" with sanctions against industrial products and other, perhaps more costly, sectors of the economy.

INTELLECTUAL PROPERTY RIGHTS

Sanctions and cross-retaliation are powerful economic instruments. In fact, the mere threat of sanctions is often sufficient to persuade countries to change their laws or other trade practices. Sanctions are also a single-sided sword, their effectiveness being relative to the disputing countries' economic dependence, market shares and import sensitivity. Some critics of the WTO have suggested that cross-retaliation was invented to enforce the new rules on intellectual property rights—perhaps the most onerous of the 28 Uruguay Round agreements for the developing countries, whose valuable agricultural and medicinal biodiversity can be easily appropriated by foreign corporations thanks to the Uruguay Round Agreement on Trade-Related Intellectual Property Rights.

Indeed, the TRIPs Agreement, as it is usually called, was conceived by the Intellectual Property Committee, whose membership reads like a laundry list of the Fortune 500. The members at the time of the Uruguay Round were Bristol Myers, DuPont, General Electric, General Motors, Hewlett Packard, IBM, Johnson and Johnson, Merck, Monsanto, Pfizer, Rockwell, and Time-Warner. As Monsanto representative James Enyart told Les Nouvelles, a French journal, "Industry has identified a major problem in international trade. It crafted a solution, reduced it to a concrete proposal and sold it to our own and other governments... We went to Geneva where we presented [our] document to the staff of the GATT Secretariat. What I have described to you is absolutely unprecedented in GATT." His summation: "The industries and traders of world commerce have played the role of patients, the diagnosticians and the prescribing physicians."

Now, just what illness were these physicians trying to heal? "Intellectual property rights"—or "IPRs" as they are called—have traditionally fallen under the domain of national law. They generally take the form of patents, trademarks or copyrights and grant exclusive monopolies over an invention or other useful knowledge for extended periods of time, usually 17-20 years or more. Different countries have defined different IPR laws, each one a balance between industry's desire to capitalize on its investments in technological development and the rights of society to benefit from its intellectual as well as its financial contributions to industry.

In other words, when geniuses like Thomas Edison or the fellow who invented the safety pin spend years of unremunerated work in their attics and basements developing something of use to society, the objective is to ensure they get their share of the reward, usually a percentage of the profits turned by the company that commercializes the product. Likewise, IPRs are supposed to guarantee a fair return to the many authors and musicians and others who produce work of cultural value to society, paying royalties on the numbers of books or records sold.

Today, however, intellectual property rights are becoming the weapon of choice of giant corporations to monopolize a product regardless of its social value. Almost daily in the business pages of the newspaper, we can read about other countries' violations of intellectual property rights. The recording industry, Hollywood, designer clothing companies and other U.S.-based transnational corporations all depend upon the U.S. Trade Representative to force China and Taiwan, Argentina , India,, Thailand and Brazil for example, to crack down on local entrepreneurs trying to make a buck selling counterfeit name-brand products that consumers crave once global advertising reaches them.

On the one hand, one can sympathize with the likes of Madonna and Levi Strauss and their aversion to having their good names ripped off. And of course, consumers should not be ripped off, either by sub-standard quality merchandise. On the other hand, why should a handful of super-rich icons and their corporate sponsors control the entire global market for music or blue jeans? Indeed, as corporate sponsorship becomes essential to successful marketing in the global economy, the benefits of cultural IPRs will concentrate in corporate coffers. An analyst for the Wall Street Journal has concluded, for example, that the enforcement of IPRs in the music industry could lead to a net outflow of money from Africa, despite the increasing popularity of African musicians, as the revenues flow to a handful of recording companies instead of to the artists themselves.

Furthermore, very few small firms and independent entrepreneurs are able to utilize IPR laws to defend their interests in the global economy. The vast majority of patent applications filed today are sponsored by corporations or other large institutions such as universities, not individuals. A patent application can cost $10,000 or more, and if it is disputed, litigation costs can run upwards of a quarter million dollars. What brilliant artists singing before local crowds or underemployed inventors conducting experiments at home can afford to defend their products, even if they were shrewd enough to get to the patent office before a corporate sponsor or copycat?

And should the U.S. taxpayer foot the bill for

enforcing these corporate rights around the world? Believe it or not, such "economic crimes" by corporations against other corporations have become a primary target of the FBI, as of October 1996, when Congress passed the Economic Espionage Act. "The nation's economic interests are a part of its national security interest. Thus, threats to the nation's economic interest are threats to the nation's vital security interests" reads the introduction to the Act, which amends the federal criminal code "to prohibit wrongfully copying or otherwise controlling economic proprietary information" with fines up to $10 million and imprisonment of up to 15 years per offense. The Economic Espionage Act of 1996 also authorizes the FBI to use wiretaps and otherwise aggressively pursue violators of intellectual property rights—using our tax dollars to protect giant corporations' trade secrets.

PATENTING PLANTS AND OTHER FORMS OF LIFE

Thanks to intellectual property rights, plant breeding is fast coming under the control of a small number of transnational companies, instead of farmers. Whereas in recent years, many thousands of mom & pop seed companies flourished around the world, they are rapidly being subsumed by the chemical industry. Of 1,500 seed companies in the world in 1995, 24 held a combined market share of more than 50%, as parent companies in the agricultural sector with subsidiaries in food processing, trade and agro-chemicals. Of the 24, eight were transnational corporations.

Agricultural patents are immensely valuable—especially for "tying up germplasm so that it works only with your chemical products," as a December 1996 report on the seed industry by investment advisory firm Dain Bosworth put it. In just the past year, for example, Monsanto has gained control over the genetic variability and the reproductive material of a lot of commercially valuable plants by acquiring:

- Agracetus, a tiny company which has claimed a patent for all the future genetically engineered products of the entire species of cotton and soybeans, at a cost of $150 million.
- Calgene, designer of the unsuccessful "FlavrSavr" tomato which was genetically engineered to delay rotting, thus allowing for long distance shipping and shelf storage. The first 49.4% of Calgene cost $30 million; the next 4.7% cost $50 million; and Monsanto is now bidding to acquire the rest.
- the Asgrow seed company, which had won a U.S. Supreme Court case against an Iowa farm family that sold a "brown bagged" second generation of Asgrow's seed to other farmers contrary to Asgrow's intellectual property rights, at a cost of $240 million.
- 40% of DeKalb Genetics, a corn-soybeans-sorghum seed company with a swine sideline, for $158 million, including a first right-of-refusal to buy another 50% when it comes on the market.
- and the biggest prize of them all, Holden Foundation Seeds, which supplies one third of all the seed corn planted in the United States, for a whopping $1.02 billion.

The application of patents to plants, seeds and germplasm—and thereby, their privatization—is particularly offensive to farmers, given that, collectively, farmers have cultivated improved varieties of useful plants for human use over some 10,000 years. Even today, farmers in most parts of the world continue to harvest seed at season's end, selecting the best for the next sowing, and sharing them with their neighbors. Yet, under the new Uruguay Round rules, these traditional agricultural practices will become illegal, once patented seeds are introduced to farming communities.

The actual text of the TRIPs Agreement states that all members of the World Trade Organization "shall provide for the protection of plant varieties either by patents or by an effective sui generis system or by any combination thereof." (Sui generis is a Latin phrase meaning "of their own kind.") Simultaneously, governments are given the option to exclude from patentability: "plants and animals other than microorganisms, and essentially biological processes for the production of plants or animals other than non-biological and microbiological processes." After sorting out the double-negatives, this means that anything that can be genetically manipulated can be patented and monopolized as the private property of giant transnational agricultural and pharmaceutical corporations. With patent legislation or an alternative system deemed "effective" by the WTO in place in every country, these companies can take over national industries easily and claim the whole world as their market.

So offensive is this measure, half a million peasant farmers in India protested against the Uruguay Round TRIPs Agreement a few years ago. The governments of Brazil and Argentina, too, have faced massive popular and political resistance, as their congresses and parliaments have faced legislative proposals that would change their national laws to bring them into conformity with the TRIPs Agreement.

Historically, the approach to this issue has been to promote domestic agricultural and pharmaceutical development by allowing patents on processes instead of products. Thus, a company could develop and market an improved seed or drug and earn royalties for the engineering breakthrough but not monopolize the end product. If another company could achieve an identical result through its own engineering prowess, why, so much the better—competition would keep consumer prices down and the public would have better access to the socially useful product. At the same time, the industry could grow and even compete in the international marketplace.

Eventually, the Brazilian Congress devised legislation that conforms with TRIPs, but leaves open an important loophole: prohibitions on the direct patenting of plants and animals and natural biological products and processes, although the genes spliced into genetically engineered plants and animals and other forms of life are patentable. The Indian Parliament, so far, has resisted changing its national law to conform with TRIPs, despite a July 1997 WTO dispute panel decision siding with the U.S. against India. Adding to the pressure, the U.S. Ambassador to India announced that "certain areas of research and training will be closed to cooperation" if India fails to amend its patent laws, threatening some 130 scientific projects supported by the U.S.-India Fund. In April 1997, completely ignoring the WTO process for dispute settlement, the United States unilaterally cancelled

$260 million worth of Argentina's trade on grounds that Argentina's intellectual property laws did not comply with "international standards." The U.S. has brought similar pressure against Pakistan, Ecuador, Thailand, Ethiopia, and other countries whose intellectual property regimes fail to satisfy transnational agricultural and pharmaceutical interests.

And so the race to genetically engineer everything and be first to the patent offices is on. The immense amount of genetic data being churned out by automated DNA sequencers has created a flood of patent applications to the U.S. Patent and Trademarks Office (PTO). According to PTO staff, it could take a single senior staffperson 90 years to examine what has already been submitted for patents. The administrative costs of processing these applications would add up to more than $20 million, yet fees would come only to about $100,000.

Once again, federal tax dollars are being spent to subsidize the private sector's worldwide acquisition of what has traditionally been a public good: germplasm, the basic stuff of life.

BIOTECH FOR PROFIT

According to Monsanto's Chief Executive Officer Robert Shapiro, "...new technology is the only alternative to one of two disasters: not feeding people—letting the Malthusian process work its magic on the population—or ecological catastrophe." Another interpretation of Monsanto's new commitment to biotechnology, however, is suggested by Dain Bosworth, the investment advisors. In their December 1996 report on recent changes in the seed industry, Dain Bosworth's analysts wrote that Monsanto's payment of $1.02 billion to acquire Holden Seeds had "very little to do with Holden as a seed company and a lot to do with the battle between the chemical giants for future sales of herbicides and insecticides."

The vast majority of scientific research being undertaken today is driven more by the goal of being first in line at the patent office—and selling chemicals—than that of meeting profound social goals. If you think this is far-fetched, ask a scientist working in a public university whether patents promote the research and development of socially useful food crops and medicines. Increasingly, more and more will answer that the institutional goal of patenting their research has forced them to work in secrecy, not publishing their findings until after the patent application has been filed. An AIDS researcher, for example, may be constrained from sharing preliminary findings with like researchers at a different laboratory in order not to give away some key secret enabling the other guys to get the patent first. Such constraints could delay the development of an effective drug for AIDS by many years.

Scientists employed by corporations have it even worse. No matter how much the biotechnology industry likes to claim that its goal is to feed the starving masses of the 21st century, corporate research is pretty well limited to products that will most immediately benefit the company. A survey by the Union of Concerned Scientists found that 93% of genetically-engineered food crops undergoing field tests are intended to make food processing more profitable, while just 7% focused on nutritional or flavor enhancement. Recently, the emphasis has been on developing plant varieties that can withstand massive doses of a company's own herbicide. This means, as the weeds get hardier and hardier, a field can be sprayed with ever-greater doses of the stuff without killing off the crop itself—which is hardly good for the health of farmworkers or those of us who eat these foods!

Take Monsanto. With a patent expiring in the year 2000 on its popular "Roundup" herbicide, which accounts for 17% of the company's nearly $9 billion in annual sales, Monsanto's researchers in the spring of 1996 unveiled their new product: genetically engineered soybeans and cotton designed to survive direct applications of Roundup. The "Roundup-Ready" crops enable farmers to spray entire fields with the herbicide, rather than applying it only on the weeds. As CEO Shapiro explains, "By replacing plowing with application of herbicides like Roundup—a practice called conservation tillage—farmers end up with better soil quality and less topsoil erosion. When sprayed onto a field before crop planting, Roundup kills the weeds, eliminating the need for plowing."

Of course, it also sells a lot of herbicide. Farmers buying the Roundup-Ready seed must sign a contract promising to use only Roundup on the crop, under penalty of paying 100 times more than the seed originally cost, and giving Monsanto the right to send its agents onto their farms at any time to verify compliance. Monsanto's major competitor, AgrEvo, has just gone to market with its own "Liberty Link" seed, genetically engineered to grow plants that can tolerate applications of AgrEvo's herbicide called "Liberty."

As these chemical behemoths use seeds, the essence of all future plant life, to fight for control of the herbicide market, a few naysayers ask: does society really need more chemicals to be used more widely on the farm? And what if the broadcast spray reaches neighboring woodlands or a neighbor's non-chemical-ready crop?

Another major project of agribusiness researchers has been to genetically engineer corn, cotton and other basic crops with genes from a natural bacteria that can kill pests. This bacteria, called Bacillus thuringiensis or "Bt" for short, is widely used by organic gardeners as a natural non-toxic pest control. But once a crop covering millions of square miles is engineered to express the Bt gene, it's only a matter of a few years before the insects develop immunity to Bt. Then, the genetically-engineered crops will be vulnerable to new super-insects and the original organic pesticide will be useless.

It is ironic that so many environmental disputes in the United States revolve around the issue of "takings"—that is, whether or not the government has the right to take land from private holders in order to protect endangered species, critical habitat or other public interests—much like the law of "eminent domain" allows the government to take private land for a highway. In seeking to protect the Spotted Owl or limit the use of national parkland for private grazing, environmental organizations have earned the wrath of the "Wise Use" movement, which is defending its own economic interests. It is interesting that the Wise Use people have managed to generate such a hue and cry against the relatively unusual event of a taking from private interests for the public interest, when so little opposition can be rallied against the transfer of public resources—be they public parks, utility services, municipal trash disposal, the administration of welfare programs, and even the United States Social Security system—to private profit-making corporations.

Might not the probable destruction of Bt's effectiveness as a natural pesticide and the contamination of neighboring fields, streams and woodlands when herbicides are broadcast be considered "takings?"

FOOD SAFETY AND CONSUMER HEALTH RISKS

In 1992, the United States Food and Drug Administration, the FDA, decided it would not require genetically engineered foods to undergo testing before going to market nor would it require any formal notification when a new genetically engineered food was brought to market. Only when a company declares that "sufficient safety questions exist"—a pretty unlikely scenario in this world -- would the FDA seek pre-market testing.

No one knows what health risks, if any, these genetically engineered food crops really raise for the consumer, since they are so new and testing is not required. However, there is certainly room for doubt. For example, scientists at the state universities of Nebraska and Wisconsin inserted a gene from Brazil nuts into a soybean plant with the goal of engineering a more perfect protein. They found that the amino acids missing in soybeans could be transferred nicely along with the Brazil nut gene. But these public university scientists did some testing not required by law and found that the transferred gene carried with it an allergen that caused a dangerous reaction in the tissues of human volunteer test subjects.

Remember the campaign to keep milk laced with genetically-engineered growth hormones and antibiotics out of our food supply? Short for "recombinant Bovine Growth Hormone," rBGH is a drug that can be injected daily into dairy cows so they'll produce more milk. Over-lactating, however, often causes udder infections in the cows, so then they're injected with antibiotics which flow into the milk and then into our children and other milk-drinkers.

Monsanto, the company that first won federal approval for commercial marketing of rBGH products in the United States, and the FDA claim there's no evidence that rBGH itself is harmful to consumers. Other scientists are not so sure. The FDA, for example, says that increases in the presence of a protein hormone called IGF-1, which stands for "Insulin-like Growth Factor-one," due to the use of rBGH are inconsequential—that it is inactive when ingested by rats and rendered obsolete under conditions used to process milk into infant formula. On the other hand, Dr. Samuel Epstein of the University of Chicago's School of Public Health says that IGF-1 "induces the malignant transformation of normal human breast epithelial cells."

Furthermore, there is already a glut of milk in this country which translates into low prices for dairy farmers, many of whom have gone bankrupt (although you'll never see this low price reflected in the dairy section of your supermarket). For all these reasons, consumers and farm groups tried to get the federal government to ban rBGH-products—or, at least, to label them so at the grocery store we have a choice.

Monsanto easily defeated these calls for a ban and labeling. For its part, the FDA opposed labeling on grounds there was no significant difference between the genetically engineered rBGH and the natural hormone for growth in a normal cow. The FDA issued rules that went so far as to require that those dairies choosing to voluntarily label their product as "rBGH-free" had to include on the label a disclaimer to the effect that "no significant difference has been shown between milk derived from rBGH-treated and no-rBGH-treated cows."

Despite the fact that rBGH-laced milk and genetically engineered corn and soybeans have been approved for commercial use in the U.S., our major customers in Europe are more cautious. As the first harvest of Monsanto's and other companies' genetically engineered soybeans and corn have come to market, European consumers were very clear that they did not intend to become guinea pigs. They announced boycotts of the major supermarket chains and thronged the ports and docks where the shipments were due to arrive, protesting their introduction into their food supply—so far, with fair results. Numerous European countries have moved to ban the introduction of genetically engineered foods or, at a minimum, to require them to be labeled.

TRADE WARS

In response to the European demand that U.S. exports of genetically engineered foods be labeled, the U.S. government and industry have claimed that the genetically-engineered products cannot be separated from the rest of the corn or soybean harvest. Countering this, Europe threatened to ban the import of the entire U.S. crop. Then it was the United States' turn to threaten trade sanctions—and so it goes in a swiftly-changing political battlefield in which consumer preferences are probably less important than the commercial value of the market. Commenting on these negotiations, a U.S. government official said, "We're talking big-ticket figures here."

Similarly, after the European Union (EU) banned the large-scale commercial use of rBGH in 1994, effective through the end of 1999, and called upon the U.S. to label its meat and dairy products using the drug. In response, the U.S. charged Europe with using consumer health as a disguise for commercial protectionism. When the U.S. took the EU to the WTO on this matter, which cost U.S. beef producers $100 million in lost sales, the dispute panel sided with the United States in July 1997, on grounds that the European Union had not conducted a scientific risk assessment. The European Union filed an immediate appeal claiming there is ample scientific evidence regarding the health hazards of artificial hormones, and that the panel decision restricts "the right of governments to decide what level of protection they consider appropriate for their citizens."

Trade disputes over food safety fall under the jurisdiction of the Uruguay Round Agreement on the Application of Sanitary and Phytosanitary Measures which recognizes an international bureaucracy called "Codex Alimentarious" as the official standard-setting body governing the health and safety of traded foods.

But, alas, Codex is tied closely to transnational food corporations. Upon its founding in 1962, the U.S. food industry financed the U.S. government's participation in Codex. In recent meetings, the official U.S. delegation to Codex Alimentarious has included executives of the Nestle Corporation, Coca-Cola, Pepsi, Hershey, Ralston Purina, and Kraft—as well as representatives of the Grocery Manufacturers of America, the Food Marketing Institute, the American Frozen Food Institute, the Food Processors Association and the Association of Cereal Chemists. During the two-year period of meetings, referred to as Codex's 19th Session, from 1989-1991, a total of 445 industry representatives served on national delegations, compared to only 8 representatives of public interest, non-governmental organizations.

Under the Uruguay Round Agreement on the Application of Sanitary and Phytosanitary Measures, or the "SPS Agreement" as it is called for short, governments with food safety regulations more stringent than Codex's standards must bring their national and sub-federal rules into conformance with the weaker standard, in order to facilitate trade. If one country—let's say a country exporting genetically engineered foods—chooses to challenge another country's more stringent law—let's say, one banning genetically engineered foods, a World Trade Organization dispute panel is instructed to determine whether the country with stricter standards can prove that those standards are:

- "necessary" for the protection of human, animal or plant life; and
- based on "scientific principles" and not maintained

"without sufficient scientific evidence"; or where such evidence is insufficient, they may "provisionally" adopt regulations based on "available pertinent information" while seeking information "for a more objective assessment of risk" and a subsequent review.

But what is "necessary" or "scientific"—and who should make such a determination? If a dispute between two countries reaches the WTO, the disputing parties must agree upon a three-to-five-person panel selected from a list of pre-approved trade experts to evaluate these claims. So already, the deck is stacked. These experts are trade experts—not food safety, health or environmental experts—and they are committed to evaluating the dispute in terms of its impact on trade. The government wishing to defend more stringent health and food safety standards bears the burden of proof. Yet where a government is less protective of health, Codex and the WTO claim no jurisdiction. Thus, the net effect of the Codex/WTO standard-setting regime is to lower health and safety regulation to a least common denominator.

In 1995, Codex began work to define "sound science" and the health risks of using hormones to promote growth in beef cattle and milk production in dairy cows, and is now developing rules on food labeling and genetic engineering. If the U.S. wins, the European consumers will become the guinea pigs gathering further scientific evidence in their livers and spleens, and possibly even in their genes, about the real health effects of eating genetically engineered milk and beef, corn and soybeans, and whatever is next. If Europe wins, however, and the abundant supply of U.S.-grown genetically engineered foods can't be exported, guess who's going to be sold this bill of goods: why, you and me—the American consumer!

BIOSAFETY

There are also risks to the environment when genetically engineered organisms—be they animals, plants, bacteria or other microorganisms—are released into nature. Danish scientists proved in 1996 that transgenic escape does occur, when a genetically engineered plant cross-breeds with wild relatives through insect or wind or other natural pollination processes. As the engineered characteristic crosses into wild varieties, its effects could alter the viability of the species or affect whole ecosystems.

Genetically engineered fish pose especially grave ecological risks, since they can move easily from the controlled environment to the wild. Of the 46 exotic fishes which have successfully invaded U.S. waters to date, for example, 22 escaped from aquaculture facilities. Salmon engineered to grow larger and faster than wild salmon, for example, could escape their pens and easily out-compete their wild relatives for food and reproductive opportunities—permanently disrupting critical links in the food chain.

Scientists at the State University of Oregon working with a bacteria genetically engineered to speed up the transformation of agricultural waste to ethanol decided to conduct some extra testing not required by law to satisfy their own curiosity about their discovery. To their horror, they discovered that the bacteria not only produces ethanol nicely, it also depletes a fungus in the soil that is essential to a young plant's ability to take in nitrogen through its root system. If these bacteria had been released commercially, who knows how many acres of crops would have died of nitrogen-deficiency and become virtual deserts?

Like the health tests administered by the state university scientists on the genetically engineered soybean/Brazil nut, the extra testing conducted on this ethanol-producing bacteria was entirely voluntary. The U.S. Food and Drug Administration merely urges testing when the experimenters themselves are "reasonably suspicious" of any risk inherent in the experimental genes. The U.S. Environmental Protection Agency and Department of Agriculture merely require companies to notify the federal government before conducting field tests for unusual modifications of corn, soybeans, tobacco, cotton, tomatoes and potatoes, although for other crops, they must at least get a permit. And genetically altered microorganisms must be registered, although lobbyists are now working Capitol Hill to exempt virus- and herbicide-resistant plants from even this minimal accountability. And there are absolutely no federal regulations governing genetically engineered fish.

So far, well over 2,000 applications for the experimental release of genetically engineered organisms have been filed with the U.S. government alone. In developing countries, many of which have no regulatory oversight in place, at least 90 releases of genetically engineered plants have been documented. Illegal releases have occurred in India, where 80 different genetically engineered species of microbes were illegally imported from Japan and released into the field, in Kenya, and in Argentina, where a genetically engineered vacinnia-rabies virus was set loose in 1986.

Out of concern for such potential threats, the 150-some governments signing the Convention on Biological Diversity at the 1992 Earth Summit in Rio de Janeiro called for the negotiation of a "biosafety protocol." (A protocol, in international legal terminology, defines the way in which the broad goals of a treaty are to be implemented.)

For more than three years, the United States—which to this date has not ratified this treaty—tried to block these negotiations. Then, after concerted lobbying by a handful of dedicated non-profit organizations and real determination by the other 150-some governments, the U.S. switched to hard-ball bargaining intended to minimize the scope of such a protocol. Let's exclude from the risk assessment process all genetically engineered corn, wheat, soybeans, squash and other crops that we insist are safe, U.S. officials say, so as not to place "unneeded burdens on international trade." Let's not even discuss the potential social and economic problems, they say, that could result from widespread use of genetically engineered organisms.

SOCIOECONOMIC IMPACTS OF THE NEW BIOTECHNOLOGIES

The United States' rejection of any need to consider the social and economic impacts of a brand-new. untested, globalizing industry may prove to be its undoing. In a move unprecedented in international negotiations of this sort, the African countries split off from many Latin American and other Third World countries during the 1996 negotiations on the biosafety protocol and refused to give in to the U.S. position.

(If you're not familiar with United Nations politics, there is not a voting procedure but decisions are made by consensus. Thus, if any one country—and especially if a large group of countries—refuses to budge on a particular item, it remains in the negotiating text. Of course, as long as there is not unanimity, it means the negotiations are stalemated. For this reason, international agreements tend to take a long time and, in the end, tend to express whatever the world's biggest bullies, usually the United States and sometimes Europe, want them to express.)

Nonetheless, in negotiating the biosafety protocol, Africa didn't budge—insisting on the broadest possible scope—because they knew that not only their ecologically-diverse environments but their whole economies were at risk. Many African countries depend upon a single export commodity for their entire flow of foreign exchange; others depend upon just four or five major exports. These commodities are essentially raw biological material, which makes up forty percent of all present production and processing in the world. Once a corporate laboratory reduces the biological material to its molecular components, it can often be manufactured instead of harvested.

Already synthetic substitutes for cocoa and vanilla are being manufactured in factories in the U.S. and Europe—resulting in huge losses in export volume as well as revenue per ton to Africa, as gluts of the real commodities accumulate and drive down prices. Just think about how many tons of sugar are no longer exported from the Caribbean nations, thanks to those of us drinking diet soda and stirring Nutri-Sweet into our coffee!

Most insidious of all, by the act of moving one gene around within a cell, a manufacturer can claim a patent for "inventing" a "novel" product and charge royalty fees on top of the sales price for a genetically engineered synthetic substitute that the Third World country once produced itself. The International Labour Organization, a body established by the United Nations, has suggested that in the near future the Third World could lose up to 50% of its jobs, due to the new biotechnologies.

Indeed, the advent of the new biotechnologies could turn farmers into factory workers everywhere. The technical possibilities of reducing agricultural products to their central components, such as fats or carbohydrates or enzymes or what-have-you, for both food and industrial purposes mean that the raw materials are increasingly interchangeable. In other words, Nebraska corn farmers could soon find themselves competing with the Japanese fishing industry and Malaysian palm oil plantations. They would all turn their harvests in to a conglomerate like ADM—the corrupt transnational convicted of price-fixing that calls itself "Supermarket to the World"—where corn, fish and palm oil alike could be transformed into the same high-tech industrial ingredient.

If Monsanto's contract with farmers who use their genetically engineered seed is any indication, the giant companies will soon be issuing competitive contracts around the world that stipulate what inputs must be used and where to deliver such-and-such volume of the product by a certain date for a prescribed price. The contractors will pledge an entire season's work, bear all the risk, and never know what they really produced.

DRIVING OUT TRADITIONAL AGRICULTURAL SYSTEMS

Biotechnology proponents claim that their research is the key to solving world hunger. They argue that the "Green Revolution" which brought super-high-yielding varieties of wheat and rice to Latin America and Asia was such a success, it is time to initiate a Second Green Revolution—especially in poverty-stricken Africa—utilizing the new genetically engineered varieties of herbicide-resistant plants in addition to the higher-yielding hybrids of earlier decades. Norman Borlaug, the Nobel Prize-winning "father" of the Green Revolution, has joined up with a former World Bank agriculture expert and, financed by a Japanese industrialist, is organizing fertilizer manufacturers and big seed companies, including the Monsanto and Cargill corporations, to expand their markets in Africa. They claim that doing so will enable agriculture to become a "powerful engine" for economic growth there.

The first Green Revolution was certainly successful, in so far as its goal was to substitute high-yielding commercial varieties of wheat, rice and other basic grains for the traditional crops of much of Asia and parts of Latin America. The first Green Revolution also substituted chemicals, tractors and combines, irrigation systems and other capital-intensive technologies for labor in these regions, sending many peasants to the slums of the cities to eke out a living as best they could.

In the Philippines, for example, the rate of rice production doubled every year from 1965 to 1980 thanks to using so-called "Miracle Rice" seed, expanding irrigated land and quintupling fertilizer use. The Green Revolution was so well-promoted, by 1990 over 80% of Filipino rice paddies were planted in one of several varieties of Miracle Rice.

Sound good? Sound like the solution to world hunger? Well, as time went on, Miracle Rice began to seem like less of a miracle. The monocultural environment became a perfect incubator for the Brown Planthopper, an insect which can ravage millions of metric tons of rice per year. New insecticides were invented to combat the pest, but, of course, the insects evolve quicker than the scientists can formulate and market each new pesticide. As Peter Kenmore, the head of the United Nations' Food and Agriculture Organization's Integrated Pest Management Program for Asia and the Pacific, put it, "Trying to control such a population outbreak with insecticides is like pouring kerosene on a house fire."

Over the years, some 4,000 Filipino rice farmers have died of pesticide poisoning and two thirds of the Philippines' rice paddies are now chemically degraded. Fertilizer use quintupled over the past few decades, and even the much-touted high yields are now in decline.

Similar reports of little mentioned side effects come in from elsewhere.

In Zimbabwe, a few years back, the World Bank encouraged the government to give peasants hybrid corn seed (or maize, as they call it there) in packages complete with chemical fertilizer, as a requirement if they wanted financial credit. They planted the maize instead of their traditional food crops, sorghum and millet, for sale with the idea that the income would allow the family to buy food and meet other needs. But by harvest-time, maize was selling for just half of what it cost to send a single child to school.

Then came the 1992 drought and food production fell by half, because the foreign maize was not drought-resistant. The maize was also vulnerable to a toxic fungi widespread in Africa, which causes liver cancer and suppresses the immune system in children. Since the Zimbabwean government, like much of the world, had also sold off its reserve supply—in this case, to pay off World Bank loans—food shortages struck hard. In the communities where the maize-packages had been promoted, childhood malnutrition became commonplace—whereas the communities still producing sorghum and millet for their own consumption at least had enough to eat. And yet, sales of the maize produced by these families, who did not even have enough to eat, showed up as "economic growth" for the Zimbabwean economy.

No doubt the next World Bank promotion will be genetically engineered maize seed, bringing Monsanto to Zimbabwe along with all of its chemicals. Since there is yet no international biosafety protocol and Zimbabwean national regulation so far simply calls for the voluntary registration of new plant varieties, the Zimbabwean peasants will join the global community of human guinea pigs upon whom genetically engineered foods are about to be tested on a massive scale.

BIODIVERSITY AND FOOD SECURITY

In Zimbabwe, the high cost of the foreign seed and chemicals and the low returns in the market made many communities suspicious of the industrialized export-oriented approach to agriculture. During the period of food shortages, many maize-producing families went to their neighbors for the loan of traditional seeds with which to rebuild their home gardens. Fortunately, most of Africa is so extremely rich in biological diversity that the capacity to rejuvenate traditional forms of agricultural production through the use of local varieties is great.

In fact, this is how agricultural biodiversity evolved in the first place. As roving clans discovered they could more easily grow food than hunt and gather, they settled down into territorial communities and learned how to cultivate plants for clothing and shelter as well as foods. They learned to collect the best seeds from each harvest to plant next and how to cross-pollinate the strongest highest-yielding plants to improve future generations. They utilized wild stock to strengthen field crops, and even traded with each other, experimentally crossing local stock with breeds cultivated elsewhere to develop new varieties that might be better suited to certain soils and water cycles.

Over time, farmers generated a vast wealth of plant genetic resources especially adapted to the particular ecological conditions of most regions on earth. In India, for example, peasants developed more than 50,000 varieties of rice. Mexican corn producers developed not only the yellow corn that we eat with relish on summertime picnics, but also white corn, blue corn, red corn, purple corn, speckled corn and more. (Next time you pick up a package of blue corn chips at the grocery store, you could offer a silent thank-you to many generations of Mexican peasants.)

Suddenly, however, in this—our own—generation, we discover that the planet's gene pool is no longer expanding. It is shrinking. The commercial emphasis on limited varieties of hybrid crops coupled with the destruction of subsistence agriculture and wilderness everywhere has created a new problem: genetic erosion. Some scientists fear that the gene pool may become too narrow for future adaptation to changing conditions like global warming, desertification, flooding and the devastation of war. As the gene pool erodes further and further, a global catastrophe could occur and famine could become widespread.

The United Nations Food and Agriculture Organization has a whole Committee on Plant Genetic Resources which concluded last year that the world had better invest in rescuing its remaining genetic resources and support the restoration of genetic diversity through on-farm programs of seed regeneration and development.

GENE BANKS AND BIOPIRATES

Alas, the rescue of existing agricultural biodiversity is easier said than done. In the first place, the trend towards industrial agriculture and the use of genetically engineered and genetically identical patented seed stock—the so-called Second Green Revolution—is accelerating, as is urbanization and the loss of wild habitat. In the second place, the world's gene banks are for the most part under-capitalized and poorly maintained.

Also called seed banks, these refrigerated warehouses are maintained by most countries for the long-term storage of a wide selection of seed varieties. There is also an international system of 16 linked gene banks coordinated by an intergovernmental body called the "Consultative Group on International Agricultural Research" or CGIAR. Generally, all of these national and international gene banks make their stocks of germplasm freely available for on-going plant breeding programs and collectively provide a kind of global insurance against potential emergency disruptions of the normal food system.

However, a 1996 report on the "State of the World's Plant Genetic Resources" compiled by the FAO found that far too much of the stored seed is dead—and that which is living is all too often uncatalogued. Even in the United States' publicly-sponsored gene bank at Fort Collins, Colorado, there is genetic erosion. By neglecting non-commercial varieties, Fort Collins has deprived the gene pool of a number of wild plants like chufa, martynia and rampion from which future foods could be adapted—perhaps after some terrible blight wipes out all popular varieties of North American wheat.

Fort Collins is a marvelous state-of-the-art facility chock full of genetic material from all over the world. Believe it or not, of the world's 20 most valuable commercial crops, only the sunflower is actually native to North America. Corn, potatoes, tomatoes and cotton come from Latin America; rice and sugar cane from Indochina; wheat, barley, grapes and apples from West Central Asia, and so on. Over the decades, the U.S. Department of Agriculture has ensured that it has access in Fort Collins to genes from everywhere.

Of course, industrialized countries and private agribusiness and pharmaceutical companies have stocked their own seed banks with the world's genes.

Indeed, the quest for wild germplasm with which to genetically engineer commercial products has resulted in a sort of contemporary "gold rush" known as bioprospecting. Ethnobotanists go to indigenous communities inquiring about their use of the local flora, sometimes offering compensation in the form of gifts or shares in any royalties that may be earned if a tip gleaned from the local healer leads to a product that can be patented and marketed.

Like gold diggers everywhere, these explorers inadvertently disrupt the indigenous communities. And once disrupted, it may be difficult or impossible for that community to restore the traditional balance between itself and the ecosystem which has sustained it, while being sustained by it. Once a little bit of the genetic material is safely stored in the bioprospecting company's gene bank, they can propagate or clone it, or develop a synthetic chemical substitute to meet all of their commercial production needs. The local community then has no control over future uses of its genetic and intellectual resources, and even the best compensation deals yield a mere fraction of the monetary value that a successful product can bring to the commercial enterprise.

Whether sponsored by governments or private industry, the removal of plant genetic resources from their on site—or "in situ"—habitats to artificial "ex situ" storage has led to great ecological, political and moral debates. Summing up these issues in 1994, the FAO's Assistant Director-General Obaidullah Khana referred to such bioprospecting as "biopiracy." The 1992 Convention on Biological Diversity admonishes the world's governments to address these issues.

Ecologically, work must be done to better sustain and develop genetic resources both in situ, in their natural habitat, and ex situ, in the gene banks and research fields. Politically, the genetically-wealthy developing countries seek compensation for and repatriation of those genetic resources pirated from their territories over the centuries as well as equal representation on the governing boards of CGIAR and other institutions now managing these resources. And morally, the Convention calls for the "approval and involvement" of and "benefit-sharing" with the local communities whose genetic and intellectual resources are proving to be so valuable.

FARMERS' RIGHTS

Indigenous and farming communities are beginning to assert their rights in various international negotiations. Several hundred representatives of indigenous peoples from around the world attended the third meeting of the parties to the Convention on Biological Diversity, for example, where the governments very tentatively considered how to implement the commitments they made to ensure the informed consent of local communities before their knowledge and resources are exploited, and how to share the benefits of that exploitation with them if the community so chooses.

Farmers and farm advocates are becoming increasingly vocal about their rights: specifically, the right to produce food, which entails access to land, water and seeds. These struggles are very active in present day Mexico, Guatemala, Brazil, South Africa, the Philippines, India and innumerable other countries. As industrialized agriculture goes global, promoted by agribusiness through the World Bank, the WTO and other trade arrangements, rural communities find themselves deprived of markets and pushed off the land. The search for oil and precious minerals, timber and grazing lands likewise leads to the seizure and devastation of vast territories—and the elimination of local residents.

To its credit, the FAO is now negotiating an international treaty called the "International Undertaking on Plant Genetic Resources," which includes terms for "Farmers' Rights." While implementation mechanisms have yet to be worked out, these rights would include the right of farmers to have access to the planet's genetic diversity for breeding purposes, and their right to be compensated for their contributions to the development of useful plants. It gets complicated, of course. How can you tell which farmers contributed to the corn, wheat, rice and potatoes we all eat every day? So the negotiators are making lists, proposing that payments for the use of the genetic material of those crops which have identifiable pedigrees from a particular country or region be settled on a bilateral basis between the country of origin and the recipient country, while payments for the use of crops with lengthy and global histories be invested in a multilateral fund and the revenues used for internationally agreed purposes.

Chief among such purposes, if Farmers' Rights are to be implemented, would be substantial support for on-farm conservation, sustainable use and development of genetic diversity. Farmers would have free access to the seeds stored away in gene banks, and gene bank managers would seek out farmers willing to participate in projects to regenerate that warehoused seed which is still viable. In addition to revitalizing existing seed stocks, farmers would also be encouraged to propagate diversity—that is, they would be given financial and technical support in terms of research, training and infrastructure to develop farmer-led breeding programs.

Then again, most farmers throughout the world work in collective fields and share their seeds and their ever-accumulating knowledge with each other freely. Thus, mechanisms to implement Farmers' Rights would need to engage communities of farmers who collectively contribute to the evolution of a valuable crop. But the United States has refused to recognize the "collective rights" of farmers and local communities. As a compromise at the FAO's meeting of the Commission on Plant Genetic Resources in Leipzig, Germany, last year, the U.S. delegation finally agreed to text acknowledging "the needs and individual rights of farmers and, collectively, where recognized by national law"—a challenge for legal scholars and advocates!

And what about patented seed? Like the FAO Commission, the parties to the Biodiversity Convention understand that intellectual property rights pose a threat to communities which have historically developed genetic resources and, at the same time, are dependent upon their continued access to these resources. After considerable debate, the Convention ended up stipulating that governments should cooperate to ensure that intellectual property rights "are supportive of and do not run counter to" the conservation and sustainable use of genetic resources.

Nonetheless, the U.S. has made it clear that it considers the TRIPs Agreement to be the final word conferring private IPRs over plant varieties and other living material. Indeed, newly-elected President Bill Clinton, back in 1993, filed an "Interpretive Statement" alongside the Convention on Biological Diversity when he submitted it to the Senate for possible ratification. In this Statement, President Clinton put the rest of the world on notice: "The Administration will therefore strongly resist any actions taken by the parties to the Convention that lead to inadequate levels of protection of intellectual property rights." With this interpretation in hand and frequent visits from agribusiness and the biotechnology industry's lobbyists, Senator Jesse Helms has successfully held up U.S. ratification of this treaty in the Senate Foreign Relations Committee for years.

It is at least ironic (if not immoral) that the United States government rejects the collective rights of human communities while rigorously enforcing the rights of corporations—which are of course made up of many individuals legally joined and collectively exonerated of their individual responsibility as persons.

ACT GLOBALLY

All official international negotiations take place between governments which recognize each other as peers. As global decision-making processes are presently structured, there is no direct democracy—merely a presumption that each national government is accountable to its people, to whom the benefits of an international agreement will trickle down. With most governments proving to be about as reliable in implementing international trickle-down as they are in spending national tax revenues on public service, a global movement of disgruntled citizens is emerging to demand some accountability.

Indeed, the first point made by Via Campesina, a young global movement of rural producers, at recent meetings of the FAO was that farmers have the inalienable right to participate in the negotiation of Farmers' Rights directly—and that the FAO had better set up opportunities for them to do so. Indigenous peoples, likewise, have pushed the parties to the Convention on Biological Diversity to co-host meetings over "benefit-sharing" and what constitutes their "approval and knowledge" regarding the use of intellectual resources of their communities, and are working with the U.N. on the adoption and implementation of a Draft Declaration on the Rights of Indigenous Peoples.

Farmers organizations, which have plagued the GATT for years, are looking toward the re-negotiation of the Uruguay Round Agriculture Agreement between now and the year 2000, while consumers are learning about Codex Alimentarious and the Uruguay Round's "SPS Agreement" on food safety.

Some of us are campaigning for a strong legally-binding protocol on biosafety, while others argue that food security is a fundamental human right and should be justiciable as such. Some want the United Nations to declare traditional and staple foods exempt from trade rules, while others want the U.N. to create a global environmental agency with powers equivalent to the World Trade Organization. And pretty nearly all of us are studying the WTO and its new assortment of (de)regulatory, judicial, legislative and executive functions.

As for the TRIPs Agreement, intellectual property rights were so controversial during the Uruguay Round that the final text says the section on the patenting of plants shall be reviewed by 1999. And so people around the world are asking themselves what a sui generis—meaning "of their own kind"-- system for protecting plant varieties might look like. Can it embrace Farmers' Rights and collective rights? Or should it merely reaffirm national sovereignty and the right of the peoples of each country to share genetic material freely? And guess what: there's an almost parenthetical exception written into the TRIPs agreement worth noting. It says that governments may exclude inventions from patentability in order to protect public order or morality, "including to protect human, animal or plant life or health or to avoid serious prejudice to the environment." Definitely, this sounds consistent with our slogan: "No Patents on Life!"

What about morality? Should living creatures—be they sheep or seeds or people—become the privatized monopolized property of a corporation, whether it be for 17 or 20 or 40 years, or forever? Is this not a form of slavery? What do the world's religions say about the private ownership of God's creation?

Have you heard of "Pharm-woman?" Believe it or not, Baylor University applied for a European patent on a human female genetically altered so that her breasts could be used as a drug factory; Baylor wanted monopoly rights over the use of human mammary glands to manufacture pharmaceuticals. How about John Moore, have you heard of him? He's a guy from Seattle who went to the hospital for a spleen operation; unbeknownst to him, his doctor patented some of his spleen cells and then ran into trouble trying to get a consent form signed after the fact. The California Supreme Court ruled that the doctor had breached his fiduciary duty by not revealing a research and financial interest in his patient's cells. And what about the attempts of the United States Department of Commerce and National Institutes of Health to patent the DNA of indigenous peoples from the territories of Panama, the Solomon Islands and Papua New Guinea? In these cases, the patent applications (and in one case, an actual patent) were withdrawn after a lot of complaining by the tribal people, the other governments and citizen activists from around the world.

What do you think about Dolly the Sheep, who was cloned? Should cloning of humans be pursued? How would you like to store a few replicas of yourself somewhere, in case you need a liver transplant one day?

Can we not make the case that malnutrition, food riots and even civil war amongst rural peoples deprived of their lands and livelihoods can result from the relentless pushing of patented seeds all over the world? And what do you think about the introduction of a cholera gene in bananas, so as to vaccinate kids with the fruit instead of a needle? What if the seed of this genetically-engineered banana somehow crosses over to wild relatives of the banana tree and cholera grows rampant? Certainly it is possible to argue that human health and the environment are affected by patents, which lead to risky experimentation, genetic erosion, monocultural cropping and intensive chemical use.

What future is before us? One of human laboratory experiments, sterile fields and robo-foods, in which the haves die of cancers while the have-nots die of starvation? Or one of ecologically and culturally diverse communities of moral persons who learn to challenge the new global regime of corporate power in the public interest?

Perhaps, in time, we can successfully insist that sanity, health and democracy prevail.

APPENDIX

Citizen activists around the world are actively campaigning to eliminate unjust international policies and institutions, reform those that are unbalanced, and promote a sound approach to the increasingly global organization of human society. Among their priorities, many activists emphasize the need for the world's governments to:

1) Implement the Marrakesh Decision on Measures Concerning the Possible Negative Effects of the Reform Programme on Least-Developed and Net Food-Importing Developing Countries promptly;

2) Acknowledge the established principle that food security is a fundamental human right, consistent with Article 25 of the 1948 Universal Declaration of Human Rights;

3) Revise the International Undertaking on Plant Genetic Resources to ensure the implementation of

Farmers' Rights, and include Farmers' Rights and other provisions of the International Undertaking as a legally-binding protocol of the Convention on Biological Diversity "to ensure that intellectual property rights support and do not run counter to" the objectives of the Convention;

4) Conclude the negotiation of a legally-binding Biosafety Protocol under the Convention on Biological Diversity with utmost speed—fully addressing socioeconomic issues and provisions for liability and compensation in the case of accidents resulting from the release of genetically engineered organisms;

5) Adopt and implement the United Nations "Draft Declaration on the Rights of Indigenous Peoples," which recognizes their right "to control, develop and protect their sciences, technologies and cultural manifestations, including human and genetic resources, seeds, medicines, knowledge of the properties of flora and fauna, oral traditions, literatures, designs and visual and performing arts" and develop protocols of the Convention on Biological Diversity consistent with these provisions;

6) Revise the Uruguay Round TRIPS clauses regarding the patenting of plants and other living material between now and the year 1999, enabling sui generis national legislation that prohibits all patents on life or that otherwise respects the rights of Indigenous Peoples, Farmers' Rights, collective community rights and the healthy functioning of genetically diverse ecosystems;

7) Revise the Uruguay Round Agreement on the Application of Sanitary and Phytosanitary Measures establishing minimum international food safety standards instead of limitations on food safety, and rid the Codex Alimentarious Commission of corporate influence;

8) Initiate negotiations toward a global Sustainable Food Security Convention to elevate food security to the highest level of priority within international law: at a minimum, enabling governments to implement national food security plans that could exempt staple foods from WTO rules and disciplines that undermine these plans; coordinating an international network of local, national and regional food reserves; facilitating international commodity agreements to ensure access to staples that nations are unable to provide for themselves; and creating mechanisms to aid governments in disputes over food and agriculture policy with other entities such as the WTO;

9) Revise the Uruguay Round Agreement on Agriculture between now and the year 2000: at a minimum, prohibiting export dumping, allowing the use of import restrictions to maintain the integrity of domestic supply management programs, and exempting from GATT rules and disciplines those products essential for food security, natural resource conservation, and the protection of human, animal or plant life or health; and

10) Build local, national, regional and global political systems based upon the full participation of all segments of civil society.

OTHER OPEN MEDIA PAMPHLET SERIES TITLES AVAILABLE